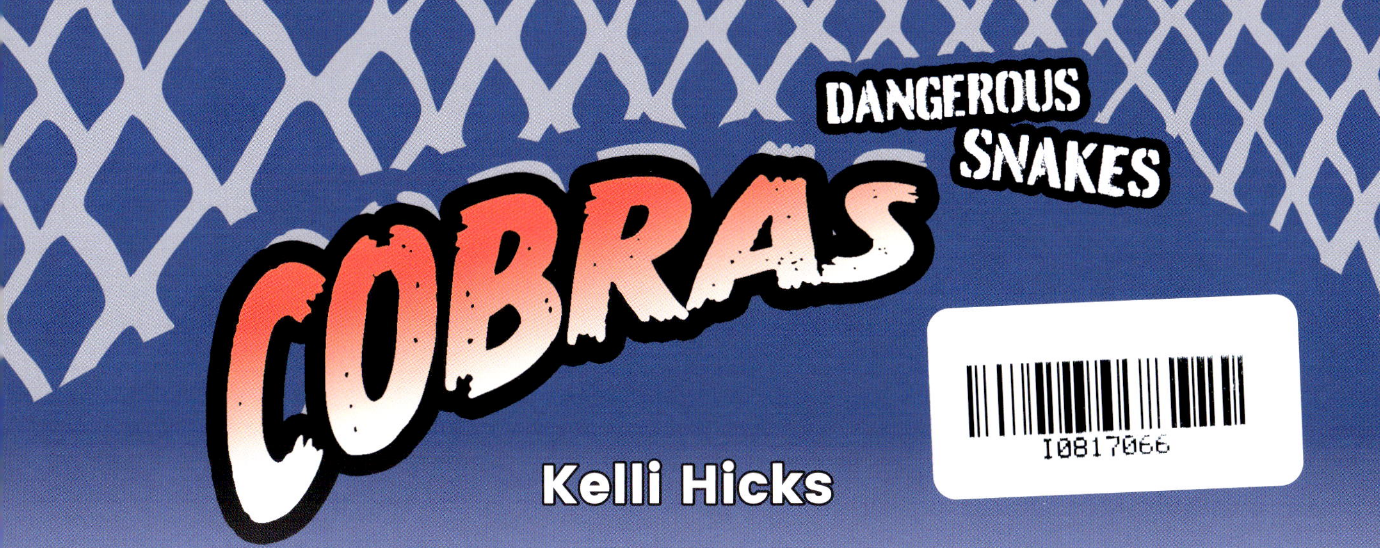

TABLE OF CONTENTS

A Crabtree Seedlings Book

School-to-Home Support for Caregivers and Teachers

This book helps children grow by letting them practice reading. Here are a few guiding questions to help the reader with building his or her comprehension skills. Possible answers appear here in red.

Before Reading:

- What do I think this book is about?
    - *I think this book is about dangerous snakes.*
    - *I think this book is about what cobras look like.*

- What do I want to learn about this topic?
    - *I want to learn where cobras live.*
    - *I want to learn what kind of food cobras eat.*

During Reading:

- I wonder why...
    - *I wonder why a cobra stretches its neck wide and flat.*
    - *I wonder how a cobra uses its tongue.*

- What have I learned so far?
    - *I have learned that cobras eat birds, small mammals, lizards, eggs, and even other snakes.*
    - *I have learned that deadly venom comes from their fangs.*

After Reading:

- What details did I learn about this topic?
    - *I learned that cobras swallow their prey whole.*
    - *I learned that cobras can go without food for a week after a meal.*

- Read the book again and look for the vocabulary words.
    - *I see the word **fangs** on page 16 and the word **tongue** on page 14. The other glossary words are on pages 22 and 23.*

COBRAS

The snake stretches its neck wide and flat.

Watch out! It is a **cobra**.

Many kinds of cobras live in the world.

Forest Cobra
6–10 feet (2–3 meters) long
Lives in Africa

Indian Cobra
4–7 feet (1–2 meters) long
Lives in Asia

King Cobra
8–18 feet (2–5 meters) long
Lives in Asia

A cobra hunts when the sun comes up or when it goes down.

The smooth **scales** on its back make a shape.

The cobra hunts its **prey**.

Cobras eat birds, small mammals, lizards, eggs—even other snakes!

It uses its **tongue** to follow its meal.

Deadly **venom** drips from hollow **fangs**.

The cobra strikes fast and swallows its meal whole.

Cobras can last weeks or months without eating another meal.

Luckily, cobras try to stay away from people.

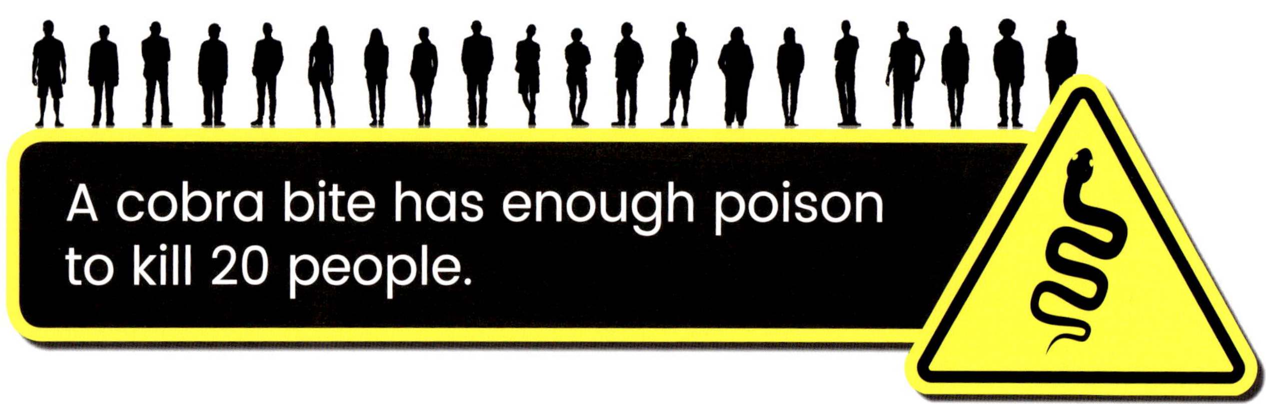

Glossary

cobra (KOH-bruh): A cobra is a large, poisonous snake that rears up and spreads its skin so that its head and neck look like a hood.

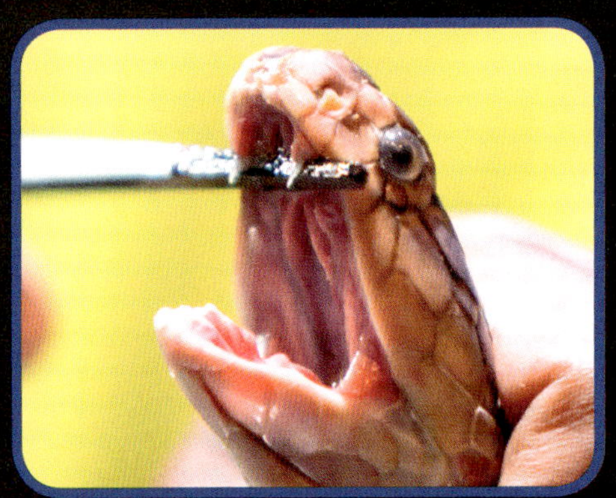

fangs (FANGZ): Fangs are sharp, long teeth.

prey (PRAY): Prey is an animal that is hunted by another animal for food.

scales (SKAYLZ): Scales are small pieces of hard skin that cover the bodies of reptiles and fish.

tongue (TUHNG): A tongue is a movable muscle in the mouth.

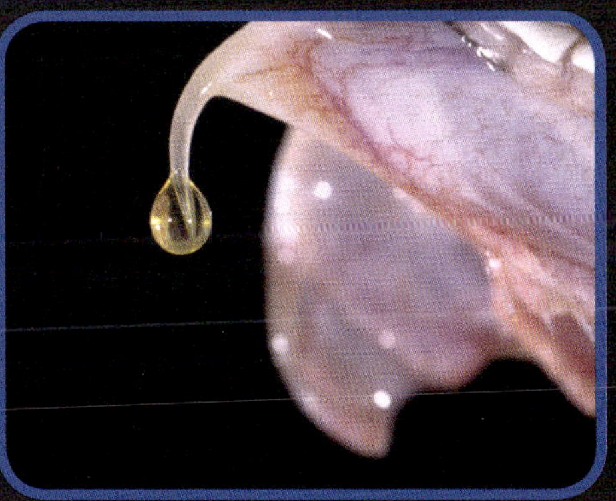

venom (VEN-uhm): Venom is a poison passed through a bite or sting.

Index

About the Author

Kelli Hicks

Kelli Hicks loves to learn about science and nature, including dangerous snakes. She prefers to read about snakes rather than meet them in person. She lives in Tampa with her husband and two children and her dog, Emma June.

Websites

https://www.livescience.com/43520-cobra-facts.html
https://easyscienceforkids.com/all-about-cobras/

Written by: Kelli Hicks
Designed by: Jennifer Dydyk
Edited by: Tracy Nelson Maurer

Photographs:
mask for snakeskin graphic on cover and pages © shutterstock.com/Merydolla; yellow triangle with snake graphic © Top Vector Studio/Shutterstock; Cover photo and pages 4 and 7 (Forest Cobra): © Skynavin/Shutterstock; page 3 © CappaPhoto/Shutterstock; page 6 © reptiles4all/Shutterstock; page 7 (Indian Cobra) © RealityImages/Shutterstock, (King Cobra) © LenSoMy/istock, (map) © Pyty/Shutterstock; page 9 © 144126157/Shutterstock; page 11 © Aleksandar Kamasi/Shutterstock; page 13 © Mufti Adi Utomo/Shutterstock; page 15 © Kristian Bell/Shutterstock; page 17 © koonsiri boonnak/Shutterstock; page 19 © 1510490567/Shutterstock; page 20 © Rawpixel.com/Shutterstock; page 21 © Stu Porter/Shutterstock; page 23 (venom) © Joe McDonald/Shutterstock

Library and Archives Canada Cataloguing in Publication

Title: Cobras / Kelli Hicks.
Names: Hicks, Kelli L., author.
Description: Series statement: Dangerous snakes | "A Crabtree seedlings book". | Includes index.
Identifiers: Canadiana (print) 20210202289 | Canadiana (ebook) 20210202297 | ISBN 9781427162335 (hardcover) | ISBN 9781427162397 (softcover) | ISBN 9781427159045 (HTML) | ISBN 9781427159052 (EPUB) | ISBN 9781427159069 (read-along ebook)
Subjects: LCSH: Cobras—Juvenile literature.
Classification: LCC QL666.O64 H53 2022 | DDC j597.96/42—dc23

Library of Congress Cataloging-in-Publication Data

Names: Hicks, Kelli L., author.
Title: Cobras / Kelli Hicks.
Description: New York : Crabtree Publishing, [2022] | Series: Dangerous snakes- a Crabtree seedlings book | Includes index.
Identifiers: LCCN 2021018641 (print) | LCCN 2021018642 (ebook) | ISBN 9781427162335 (hardcover) | ISBN 9781427162397 (paperback) | ISBN 9781427159045 (ebook) | ISBN 9781427159052 (epub) | ISBN 9781427159069
Subjects: LCSH: Cobras--Juvenile literature.
Classification: LCC QL666.O64 H53 2022 (print) | LCC QL666.O64 (ebook) | DDC 597.96/42--dc23
LC record available at https://lccn.loc.gov/2021018641
LC ebook record available at https://lccn.loc.gov/2021018642

Crabtree Publishing Company
www.crabtreebooks.com 1-800-387-7650

Printed in the U.S.A./062021/CG20210401

In Canada: We acknowledge the financial support of the Government of Canada through the Canada Book Fund for our publishing activities.

Published in the United States
Crabtree Publishing
347 Fifth Avenue, Suite 1402-145
New York, NY, 10016

Published in Canada
Crabtree Publishing
616 Welland Ave.
St. Catharines, Ontario L2M 5V6